I0750685

KENT SCIENTIFIC INSTITUTE.

(Miscellaneous Papers—No. 3)

Notes upon the Fossil Remains of the Lower Carboniferous Limestone Exposed at Grand Rapids, Mich.,

BY E. A. STRONG.

[The following remarks upon the leading fossils found in place at Grand Rapids, Mich., have been printed solely for home use and with a view first to stimulate and guide local collections, and secondly to furnish the basis of some observations upon the co-ordination of our strata with the well-known western divisions of the Lower Carboniferous rocks. To accomplish the former of these objects, it seemed desirable to determine, as far as possible, label, and display our more common fossil species; to draw up and publish a catalogue of the same, with such notes as would be useful to collectors; and to describe, under local and provisional names, such forms as could not be brought within the limits of recognized species. The author is aware how severely and how justly work of this last-named sort is condemned by systematists as irregular and confusing, but he would urge that it is not les a source of confusion to include really heterogeneous material under one term, or to extend the application of a specific name without at the same time extending the diagnosis of the species. He would also urge that fairly well-preserved fossils are rare in our rocks, that a collector sees certain characteristics in a specimen which can never be made to appear again, and especially that the effects of distortion upon fossils can be appreciated only by one who has seen them in place. As an example of this last-named source of error, No. 47, of the following list is often found with the axis of the shell at right angles to the plane of stratification, in which case the length is often hardly greater than the height of the form, although no evidence of distortion can be detected in the fossil itself.

The author purposes to continue this local list in another paper, and to follow the whole with some notes upon the stratigraphical geology of Kent County.

All measurements which occur in the following list are given in millimeters.]

Fish teeth and spines are found sparsely scattered through all the strata above the "geode bed."

Nos. 1-4. These are our more common forms of fish teeth. No. 4 appears to be *Helodus crenulatus.* Besides the above, *Psammodus*-like teeth have been observed, though most of our specimens can be referred to the genus Helodus. Nos. 2 and 4 have been found associated with No. 9.

No. 5. *Cladodus irregularis* (n. sp.):—Teeth of medium size, with a somewhat thin and narrow base more than usually irregular in surface and outline, surmounted by six cones, the height of the central one being a mean between the length and breadth of the base. Base irregularly convex in front (almost straight for the middle third) sloping backward to the bevelled and rather thin posterior border, which is irregular in outline, broadest on the side of the tooth which has fewest denticles, and having a slight sinus opposite the middle cone. The lower surface is slightly excavated beneath the median cone, and more strongly in the central portion and at the ends. Median cone nearly straight, very slightly bent forward at base, then backward, then forward at tip; upper and lower portion elliptical in section, the middle portion somewhat flattened in front; surface irregularly striated with four to six rather sharp striæ to the mm., which are less strongly marked and finally disappear at the base, the two striæ at the *sides of the middle portion* being very prominent. Lateral cusps conical, overhanging the anterior margin, three upon one side and two upon the other, turned forward, except the large exterior denticle upon one side which is strongly turned outward and backward, the corresponding external denticle upon the other side wanting in all specimens, and, apparently, not accidentally.

Three specimens found from No. 8, Taylor's quarry, one of which yields the following measurements: Total height, 29; height of median cone, 24.7; height of external denticle, 4.9; of others, about 2; average thickness of base, 3.6; length of base, 31.4; breadth of base, 17.8.

No. 8. This has been referred to *Ctenacanthus gracillimus*, N. and W., which it strongly resembles, except in the generic character, "pectination by transverse scales or tubercles."

No. 9. Fragments of a stouter, straighter form which is more common than No. 8, near Leonard street bridge and Taylor's quarry.

No. 10. *Ctenacanthus* (*?*)—Spine thick, stout, nearly straight, 99.4 long, of which about 66.4 is exposed; 13.5 wide at base; 12.4 at the middle, and 7.9 at the distance of one-fourth the length from the end, which is obtusely rounded. Greatest thickness in the middle of the exposed portion, 5.5 mm; width of implanted portion, 7.2 at middle, the end being rounded and broader and thicker than the other extremity; anterior margin, sharp, exposed portion slightly convex in outline, and implanted part concave; posterior margin nearly straight in outline with a conspicuous sulcus about 2.5 mm. deep at the middle, bordered by ridges which bear numerous minute rounded or subconical tubercles, directed somewhat downward, about 1 mm. apart, the breadth of base and height of tubercle being each about 1-8 mm.; surface, convex, with unequal, obscure, longitudinal ridges which are either persistent or disappear irregularly and increase by implantation or division, upon and between which are finer striæ, giving a fibrous appearance to the whole. In its straight lower margin and triangular section this spine resembles *C. triangularis*, Newb., but it is notably different in other respects.

Glabellæ and pygidia of a species of *Phillipsia* apparently distinct from the one described below are found in the lower strata of this formation, but are in too poor condition for identification. The following species is described from three nearly complete individuals from Scribner's quarry and the river below Leonard street bridge, and a few fragments from Taylor's quarry.

No. 14. *Phillipsia longispina*, (n. sp.)—Outline elongated elliptical, sides nearly straight, ends evenly rounded; head, thorax and pygidium nearly equal in breadth.

Glabella with posterior lobes small, anterior moderately large, evenly convex, without margin; facial suture nearly as in *P. Portlockii*; neck segment about as wide as the thoracic, and continued backward in a narrow spine which extends beyond the thorax and is applied so closely to it as not to interfere with the elliptical outline; neck furrow shallow, curving backward strongly and terminating at the lateral furrows of the cheeks. Thorax and pygidium much as in *P. Portlockii* except that the border of the latter is very broad, equaling in breadth the lateral lobes. One specimen from Scribner's quarry yields the following measurements: Length, 44.4; of head, 15.3; of thorax, 12.1; breadth of head, 21.2; of thorax, 21.9; of pygidium, 20.4.

No. 18. *Nautilus disciformis* (?) M. and W.—Two specimens, one badly weathered from a cellar—West Side—the other a fragment from a bed of Coldbrook, have been referred provisionally to this species.

No. 20. *Nautilus* sp. ?—Fragments of three individuals showing the chambered portion, from Taylor's quarry, of the type of *N. latus* and *N. Winslowi*, M. and W., but probably specifically distinct. Dorso-ventral diameter at first septum, 22 mm.; transverse diam, 53 mm.; septa distant about one-tenth the trans. diam; a broad obscure dorsal sulcus. Has been known locally as *N.* (?) *latissimus*.

No. 22. *Nautilus ellipticus*, (n. sp.)—Much resembling *N. Forbesianus*, McChesney, and *N. Spectabilis* M. and W., but having an aperture almost truly elliptical, one and a half times as high as wide, the section becoming more circular as the diameter decreases and expanding very rapidly as it passes from the septate to the non-septate portion, which latter is not nodose. This species is not rare at Taylor's quarry; is associated with zaphrentis spinulifera.

No. 23. *Nautilus Kentensis*, (n. sp.)—The form locally known by this name cannot be referred to any species known to me. The last whorl is much like *N. Niotensis*, M. and W., but as the shell expands the dorsal region becomes more prominent and sharply curved, which with the flattened ventral side gives a triangular appearance to the section; transverse and dorso-ventral diameters subequal; breadth of chambers, one-fourth to one-fifth the diameter. Same locality as above, and possibly a distorted form of the preceding.

No. 25. *Orthoceras* sp ?—Fragments of questionable identity, Coldbrook.

No. 30. *Bellerophon* sp ?—Somewhat resembles *B. ellipticus* McChesney, from the coal measures, but has a less expanded aperture and less prominent ridges; rarely found at all localities above the upper dam.

No. 31–32. Casts of two or three other species are associated with the above, one of which resembles *B. nautiloides*, Winchell, from the Marshall group, except in its larger size.

No. 33. *Porcellia nodosus*, Hall.—Two fragments of questionable identity with the above, Taylor's quarry.

No. 34. *Euomphalus rugosus*, Hall.—Casts and ill-preserved fragments are found at nearly all localities, not in a condition for identification. A few which partially retain the shell seem to me not distinguishable from the above coal measure species.

No. 35. *E. latus*, Hall.—Taylor's quarry, lower beds.

No. 36. *E. planodorsatus*, M. and W.—I have a memorandum concerning a specimen of unknown locality agreeing with the figure and description of the above species, but I cannot now find the specimen.

No. 37. Casts of a small, moderately elevated shell of three (?) whorls sometimes weather out of the "arenaceous layer" at Wells' and Taylor's quarries.

No. 38. This form has been referred to *Pleurotomaria Chesterensis*, M. and W., but closer examination shows that it does not belong to that species if, indeed, it is a *Pleurotomaria* at all. No perfect specimen has been seen, though the fragments found show the surface characters very well, which somewhat resemble those of *Pleurotomaria subsinuata*, M. and W. Taylor's quarry.

No. 45. *Allorisma sinuata*, McChesney.—Rare, especially at Taylor's quarry; more common about the upper dam. Our specimens would be described as *moderately* instead of *strongly* sinuate.

No. 46. *Allorisma elongata*, (n. sp.)—Like the above, except more elongated posteriorly, wholly without sinus upon the ventral margin or depression upon the valves, and beaks nearer the anterior end. An undistorted specimen of medium size gives the following measurements: Length, 64.; height to hinge line, 23.5; height to summit of beaks, 25.4; greatest thickness, 20. Beaks one-ninth the length of the shell from the anterior end (varying in different specimens from one-seventh to one-tenth); twenty-eight concentric ridges—which in this case are pretty persistent—can be counted upon each valve. A specimen one and one-half inches long gave analogous results.

No. 47. *Allorisma quadrata* (n. sp.)—Like the above, except smaller, relatively broader, with posterior end more quadrate. A specimen of full size, and *undistorted*, gives the following admeasurements: Length, 31.; height, 15.8; of beaks, 16.3; height at 5 mm. from posterior end, 15.; thirty-nine concentric ridges in fasciculi of two, three, or four, which often unite upon the posterior end.'

[The author has recently endeavored to refer to the above species described several years since, about seventy specimens, collected this season. He finds none of No. 45, fourteen of No. 46, forty-one of No. 47, and a remainder of heterogeneous material including a few individuals strongly resembling Owen's figure of *A. regularis*; a few not unlike *A. clavata*, and almost every degree of variation between these forms and the three described above Many resemble Fig. 2, Pl. 22, Vol. V, Pal. Ills. He had formerly believed that Nos. 46 and 47 were distinct from No. 45, as he had never found them in the same horizon; but recently he has found the two former associated in strata 3 and 4, Taylor's quarry. It is questionable whether we have more than one species of this genus.]

No. 48. Rare and showing few distinctive features. Perhaps a *Schizodus*: associated with *zaphrentis spinulifera*.

No. 49. Fragments probably of an aviculopecten. Locality unknown.

No. 50. This is our most abundant fossil, although its range, and the area over which it is found are very limited. It occurs abundantly upon the surface of No. 6, Taylor's quarry, new excavation—part of No. 4, old excavation—literally covering this stratum wherever it has been laid bare,

but is found at no other point. The valves are usually found separated, and bear evidence of having been beach-worn previous to fossilization, so that it seems impossible to determine the species with certainty. I have been in the habit of referring this form to *Myalina*, but it can hardly belong to that genus. Small specimens resemble *Orthonota*, in having a straight hinge line, plaited dorsal region and nearly parallel margins, but adult forms are more produced and spreading posteriorly.

www.ingramcontent.com/pod-product-compliance
Lightning Source LLC
LaVergne TN
LVHW020641110826
845149LV00004B/1309
* 9 7 8 1 4 1 8 1 9 0 5 5 2 *